Earth's Changing Islands

by Neil Morris

Raintree

www.raintreepublishers.co.uk
Visit our website to find out more information about **Raintree** books.

To order:
 Phone 44 (0) 1865 888112
 Send a fax to 44 (0) 1865 314091
💻 Visit the Raintree Bookshop at **www.raintreepublishers.co.uk** to browse our catalogue and order online.

First published in Great Britain by Raintree, Halley Court, Jordan Hill, Oxford OX2 8EJ, part of Harcourt Education.
Raintree is a registered trademark of Harcourt Education Ltd.

Editorial: Nick Hunter and Catherine Clarke
Design: Michelle Lisseter and Bridge Creative Services Ltd
Picture Research: Maria Joannou and Liz Eddison
Illustrations: Bridge Creative Services Ltd
Production: Jonathan Smith

Originated by Dot Gradations Ltd
Printed and bound in China by South China Printing Company

ISBN 1 844 21395 1
07 06 05 04 03
10 9 8 7 6 5 4 3 2 1

British Library Cataloguing in Publication Data
Morris, Neil
Earth's Changing Islands. – (Landscapes and People)
551.4'2
A full catalogue record for this book is available from the British Library.

Acknowledgements
The publishers would like to thank the following for permission to reproduce photographs: Corbis pp. **7** (Michael St Maur Sheil), **12** (NASA), **13** (Owaki-Kulla), **16** (Tim Thompson), **18** (Gail Mooney), **21** (Macduff Everton), **23** (Jose F. Poblete), **25** (Jan Butchofsky-Houser), **27** (Richard Klune); Natural Science Photos pp. **11** (C Dani-l Jeske), **14** (Pete Oxford), **24** (Ben Stobart), NHPA (Any Rouse) p. **17**; Oxford Scientific Films pp. **5** (Scott Winer), **6** (Tony Martin), **9** (Hjalmar Bardarson), **15** (Robin Bush).

Cover photograph of North Male Atoll, Fauholhu Fushi, Maldives, reproduced with permission of Pictures Colour Library.

The publishers would like to thank Margaret Mackintosh for her assistance in the preparation of this book.

Contents

Any words appearing in the text in bold, **like this**, are explained in the Glossary.

What is an island?

What do you think of when you think of an island? Does it have sandy beaches and warm blue seas? Or is it a cold, rocky place battered by ocean waves and the wind? Whichever one you think of, you're right. An island can be both of these things, and many others too.

An island is a piece of land that is completely surrounded by water. The biggest islands are in the world's oceans and seas. Many smaller islands can be found in lakes and rivers. Very small islands are called **islets**.

Greenland is the biggest island in the world. It lies between the North Atlantic and Arctic oceans, off the coast of North America.

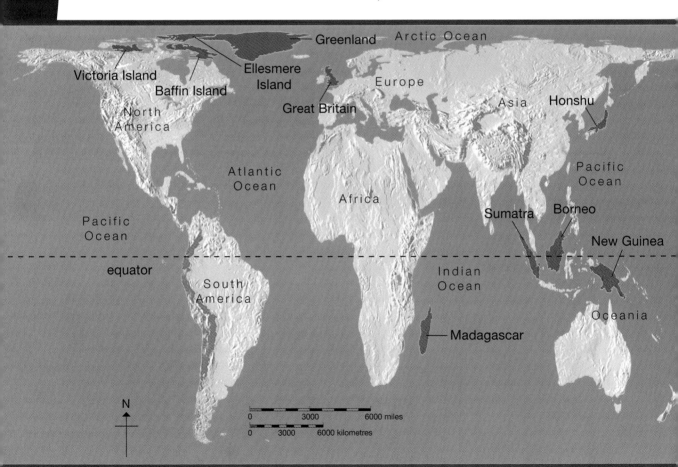

This world map shows the location of the ten biggest islands (in red). Three of them – New Guinea, Borneo and Honshu – lie in the world's largest ocean, the Pacific.

The huge land masses of Australia and Antarctica are also surrounded by water, so they are really massive islands. They are so big, though, that people call them **continents**.

Changing islands

There are many different kinds of islands and they were formed in diffferent ways. Some were made by **erupting volcanoes**. Others were formed from the skeletons of dead sea creatures. These islands are home to millions of amazing animals and plants, some of which are found nowhere else. People, too, live on, use and change islands. In recent years **tourists** have had a great effect on many of the world's smaller islands. In some places this has brought problems, which you can find out about later in the book.

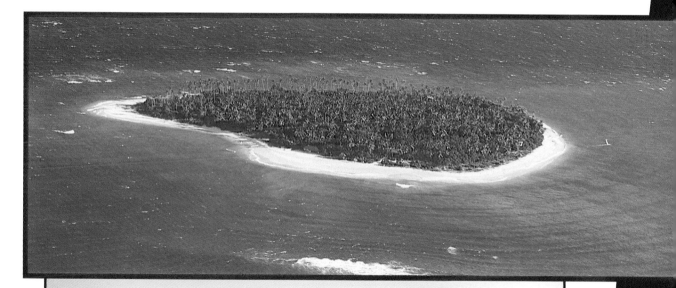

● *Islands are pieces of land that are completely surrounded by water. This is one of the many small islands that make up the country of Fiji, in the Pacific Ocean.*

World of islands

There are islands all over the world. Some, such as Madagascar, off the coast of Africa, form a separate country. Other islands are part of another country. Sumatra, for example, belongs to Indonesia. When islands are grouped closely together, they form a cluster called an **archipelago**. Japan is an archipelago made up of four large islands and hundreds of small islets.

How are islands formed?

Islands may look like any other land, but they are all really the tops of hills or mountains that rise above sea level. The rest of the hill or mountain is under water and makes up the **seabed**. Islands change slowly, over many years. The way in which islands change depends on how they were formed.

Continental islands

Islands which are close to a much larger mass of land are called **continental** islands. The mass of land is a **continent**, also called the **mainland**. Continental islands were once connected to the mainland. Some were cut off by the movement of rocks beneath the surface of Earth. This surface is called Earth's **crust**. The crust is not one continuous layer, like the shell of an egg, but is cracked into huge pieces. We call these pieces **plates**, and though they seem still to us, they actually move very slowly. As they do so, some pieces of land break away from others, and the gap between them is filled by sea. This is how the large island of Madagascar was formed, off the mainland of the continent of Africa. The island split away tens of millions of years ago, long before there were any people living on Earth.

● *Today, Greenland is almost completely covered by an ice cap that is more than 3000 metres (2 miles) thick in places. This was how most of northern Europe looked during the last ice age.*

Melting ice

Other continental islands were cut off thousands of years ago because of changes in the sea level. The British Isles are a good example of this. They were once part of mainland Europe. More than 10,000 years ago, during the last **ice age**, huge sheets of ice stretched for thousands of miles across northern Europe. When the ice age ended and large parts of these **ice sheets** started to melt, the extra water made the sea level rise. Water filled the low land between present-day France and England, forming the English Channel. Great Britain became an island.

Changing with the tide

Twice a day, the water in the world's oceans rises and then falls again. The rising and falling are called **tides**. When water covers more land, we say that the tide has come in, or that it is 'high tide'. Tidal islands are pieces of land that are cut off from the mainland at high tide. Mont-Saint-Michel is surrounded by **sandbanks**, which connect the island to the coast of northern France at low tide. When the tide rises, the sandbanks are covered by water.

● *Mont-Saint-Michel is a tidal island, about 900 metres all the way around. According to legend, a French bishop was visited in a dream by an archangel and told to build a church on the island. Its abbey was founded in 966, and the island was named after the archangel (Saint Michael in English). Today, Mont-Saint-Michel is very popular with visiting **tourists**.*

Oceanic islands

Many of the world's islands were never connected to a **continent**. Some of them lie thousands of kilometres away from the **mainland**. These islands are formed by the action of underwater **volcanoes**, which occur at the edge of Earth's **plates**. **Lava** and ash build up on the ocean floor as the volcanoes **erupt**. This forms a mountain, which keeps on growing higher, until eventually it reaches above sea level and becomes an island. Volcanic island groups such as the Philippines and Indonesia, in south-east Asia, form in regions where one plate is forced beneath another. Some volcanic **archipelagos**, such as the Aleutian Islands of Alaska, USA, have the same curved shape as the plates where they form. These are called island arcs.

● *Mauna Loa, on the island of Hawaii, is the world's largest active volcano. Nearby Mauna Kea is 36 metres higher but has not erupted for many years. A younger, smaller active volcano is called Kilauea. Another, called Loihi, is growing under the sea to the south and will eventually become an island.*

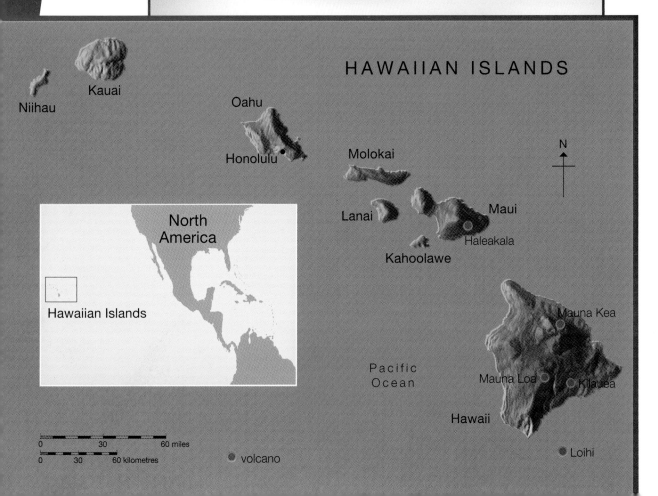

HAWAIIAN ISLANDS

Kauai

Niihau

Oahu

Honolulu

Molokai

N

Lanai

Maui

Haleakala

Kahoolawe

North America

Hawaiian Islands

Mauna Kea

Pacific Ocean

Mauna Loa

Kilauea

Hawaii

0   30   60 miles

0   30   60 kilometres

● volcano

● Loihi

Over the hot spot

The islands of Hawaii lie in the middle of the Pacific Ocean. They are a chain of eight main islands and many other **islets**. They were formed as the ocean floor moved very slowly over a hot area, called a **hot spot**, beneath Earth's **crust**. This hot spot spews out very hot or **molten** rock on to the **seabed**. Over thousands of years, layers of rock built up to form an underwater mountain. Eventually the mountain poked above the ocean surface to make an individual Hawaiian island.

An island is born

One day in 1963, fishermen saw steam rising from the sea off the coast of Iceland. At first they thought it must be smoke from a burning ship, but it was actually a volcanic island nearing the surface. By the next day, it had formed a small island, and it kept growing over the next 3 years. The island was called Surtsey. Seeds soon reached the island, carried there by the wind and by birds. They produced plants, and before long seabirds were nesting among the rocks. Over the years breeding **colonies** of birds began to appear on Surtsey.

● *The new island off Iceland is called Surtsey. It was named after the ancient Norse god Surt, who ruled over a land of fire giants.*

Coral islands

Coral islands form and grow in warm, shallow water. They are made of tiny, colourful sea animals. The animals are related to sea **anemones** and jellyfish, and are called coral polyps. Millions of these creatures live together in **colonies**. When they die, their limestone skeletons remain, and eventually become joined together. This is how coral **reefs** and islands grow. Over many years sand and soil build up on the island, so that plants can grow there. This means that people can live there, too. The Maldives in the Indian Ocean are made up of more than a thousand small coral islands. People live on more than 200 of the islands.

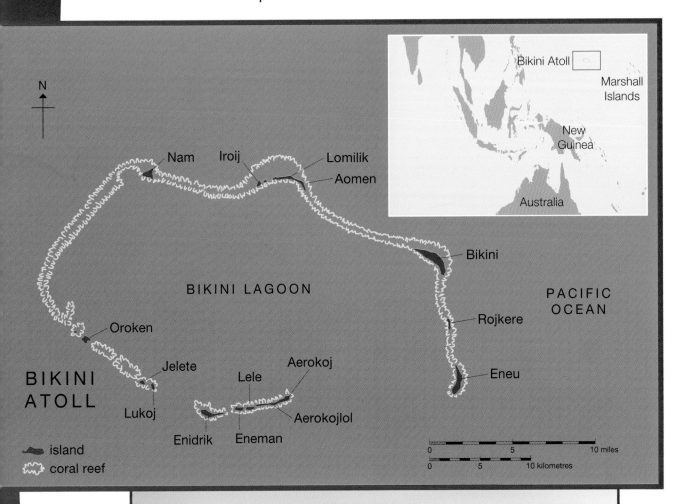

● *Bikini Atoll belongs to the Marshall Islands. Its lagoon is about 40 kilometres (25 miles) across, and its largest island is also called Bikini. From 1946 to 1958 it was used by the USA as a test site for **nuclear explosions**, and people living there had to move to another island.*

The Marshalls

The Marshall Islands, in the central Pacific Ocean, are made up of 29 **atolls** and 5 main islands. An atoll is a ring-shaped chain of coral islands. These begin as parts of a coral reef growing around a volcanic island. This is called a fringing reef. After the **volcano** has stopped **erupting**, over time it is gradually eroded (worn away). The volcanic island begins to sink and the coral forms a barrier reef. When the island has completely disappeared, the coral islands remain as an atoll. The islands surround an enclosed stretch of water called a **lagoon**.

Bora-Bora

The tropical island of Bora-Bora belongs to the Society Islands and is part of French Polynesia, in the South Pacific. It pushed up from the Pacific Ocean about 3 million years ago as a volcano. A fringe of coral grew around it, and now only the tip of the volcanic mountain remains. In time, Bora-Bora will become an atoll, like Bikini.

● *The island in the foreground (at the front of the picture) is part of the reef around Bora-Bora.*

Barrier islands

Barrier islands form near shorelines. They are made up of sand, mud and tiny stones. Ocean waves and strong winds push these materials and pile them up into a series of **ridges**. The ridges run parallel to the shore of the **mainland**, and they are sometimes called barrier beaches.

Barrier islands are very low, and sometimes they are covered with sand. In severe storms, waves carry huge amounts of sand over a barrier island, wearing it away on one side and building it up on the other. In this way, smaller islands can move towards the shore of the mainland over hundreds of years.

Long Island

Long Island (seen below from the Space Shuttle Columbia), off the coast of New York, is another kind of barrier island. It is made of rocks that were broken up, moved and laid down by **glaciers**. At the end of the last **ice age**, the rocks formed ridges. These have been worn down over the past 10,000 years, leaving a low, hilly island that is 190 kilometres (118 miles) long. Long Island also has its own sandy barrier beaches. The best known are Fire Island and Long Beach.

Hatteras Island

There are many barrier islands on the eastern, Atlantic coast of the USA, and along the gently sloping shores of the Gulf of Mexico. Hatteras Island, off the coast of North Carolina, is part of Cape Hatteras National Seashore. This coastline is known for its windswept **sand dunes** and **sandbanks**, which have always been very dangerous to shipping. Because of the number of shipwrecks, this region is known as the Graveyard of the Atlantic.

● *Boats and ships rely on Cape Hatteras Lighthouse to avoid being wrecked on the coastline.*

Changing life

The animals and plants that live on a particular island or group of islands may be cut off from other types of wildlife. Long ago, birds flew across the sea and other animals swam to the islands. Plant seeds floated across the sea or were carried by the wind or by birds. Storms sometimes carried animals and plants long distances. They adapted to their new surroundings and changed in ways that helped them to live on their particular island. Because of this, some islands have their very own kinds of wildlife.

Madagascar

As far as its plants and animals go, the large island of Madagascar really is a world apart. Tenrecs, which look like hedgehogs, and monkey-like lemurs, live only on Madagascar and the nearby Comoros islands. Two-thirds of the world's chameleons – lizards that can change colour when they are angry or frightened – also live there. More than a hundred **species** of bird are **unique** to Madagascar. The island has at least 10,000 different kinds of flowering plants, and three-quarters of these grow nowhere else on Earth.

● In Madagascar, this group of ring-tailed lemurs is sitting in the trees where they can feed on fruit, leaves and bark. Their ringed tails are longer than their bodies.

Giant tortoises

The Galápagos Islands lie in the Pacific Ocean, about 1000 kilometres from the South American **mainland**. They are home to a number of unique animals, including giant tortoises. There are several different kinds on the islands. The largest tortoise has a shell up to 1.2 metres long, and weighs about 270 kilograms. Another kind of giant tortoise, though slightly smaller, lives on the **coral atoll** of Aldabra, in the Seychelles islands of the Indian Ocean.

Land of the kiwi

The islands of New Zealand split away from an ancient **continent** about 80 million years ago. Large land animals never developed there, but there is one representative of the age of the dinosaurs. The tuatara is a lizard-like reptile that scientists believe is almost exactly the same as reptiles that lived around 200 million years ago, when dinosaurs roamed the Earth. Tuataras still live on small islands off New Zealand. There are also two famous flightless birds that live only in New Zealand. The first is the kiwi (see photo, right), which has become the national emblem. The second is the largest parrot in the world – the kakapo, or owl parrot.

Introducing animals

In the 19th century, European settlers took many animals with them in their boats when they sailed to New Zealand. These included rabbits and weasels, which soon became pests. They damaged crops and other, smaller, animals. The settlers also introduced English breeds of sheep, which farmers raised for their meat and wool. These are still important products for New Zealand today.

Island reserves

Small islands with no people act as natural **reserves** for animals and plants. Wildlife adapts (or changes) to the kinds of plants and smaller animals available to eat. Smaller animals, such as the flightless birds of New Zealand (see page 15), can survive if there are no large **predators** about to eat them. Animals and plants adapt to the local conditions.

On the Galápagos islands, there are more than ten different kinds of finches. Some of these songbirds live in the trees and feed on insects. Others spend most of their time on the ground, feeding on fallen seeds and fruits. Many of these island creatures would find it difficult to survive outside their **unique** environment. On many **inhabited** islands, the people have set up reserves, so that the plants and animals are not disturbed.

● *The US Virgin Islands National Park was set up in 1956. The waters surrounding the three main islands are included within the park. This helps protect the beautiful coral* **reefs** *that teem with colourful tropical fish. Tourists enjoy the islands' many beaches.*

Philippine eagle

The Philippines is made up of more than 7000 islands. Only about 900 of them are inhabited, but still some of the islands' animals are dying out because they no longer have enough space. The Philippine eagle, which is one of the world's largest birds of **prey**, likes to live in large forests. There, the eagle can find plenty of small animals to swoop down on and kill. It is one of 172 kinds of birds that live only on these islands. On the second largest island of Mindanao, a special centre has been set up to teach people about the eagles and to help protect them.

Protecting the 'man of the forest'

Borneo is the world's third largest island. Its tropical climate and **rainforests** make the ideal home for the orang-utan (shown below). The name of this large ape means 'man of the forest' in the Malay language. Orang-utans live only on Borneo and in a small region of the neighbouring island of Sumatra. Because much of the Borneo rainforest is being cut down (so people can use the timber), special reserves have now been set up to protect the orang-utan and other large animals, such as the Asian rhinoceros.

Changing settlements

As humans began spreading out around the world, they hunted wild animals and gathered roots and berries for food. They followed rivers and coastlines, but could not visit faraway islands until they had developed ocean-going boats. This meant that many of the world's islands were the last places on Earth to be explored and settled on. Some people, however, realized that islands had certain important advantages. The sea offered protection because islands were difficult to reach.

Crete

The Minoan **civilization** developed about 4500 years ago on the island of Crete in the Mediterranean Sea. The first people to settle in Crete probably sailed from Asia Minor (present-day Turkey) in about 7000 BC. The early Cretans were farmers, growing grain and raising sheep and goats. Around 3000 BC they discovered how to make bronze. Legend says that under the rule of King Minos, they built great palaces, such as Knossos. We call these people Minoans, after their king.

The circle of islands

The Cyclades are a group of around 30 Greek islands, in the Aegean Sea to the north of Crete. The islands got their name from the Greek word for 'circle', because they form a circle around the small island of Delos.

● *The Minoans built the great palace of Knossos, on the island of Crete, around 4000 years ago. Today, you can visit the restored ruins.*

The earliest settlement that has been uncovered is on the island of Kythnos, and may be up to 9000 years old. The first people to settle there also came from Asia Minor.

Polynesians

The people known as Polynesians live on many small islands in the Pacific Ocean. They were the first people to live on Easter Island to the east (where they settled around AD 400), Hawaii to the north (from about AD 600), and New Zealand to the south (where they landed in about AD 1000). Historians believe that the Polynesians sailed in ocean-going canoes from the mainland of south-east Asia.

Viking explorers

The Vikings were a great seafaring people of Scandinavia. They used islands as stepping stones, to travel all the way across the Atlantic Ocean. First they sailed in their longships from Norway to the Shetlands, north of Scotland. They went on to the Faroes, before settling in Iceland and then Greenland. Around 1002 a Viking named Leif Eriksson sailed even further. He named the first land he and his sailors reached Helluland, meaning 'land of flat stones'. This was probably Baffin Island, in present-day Canada.

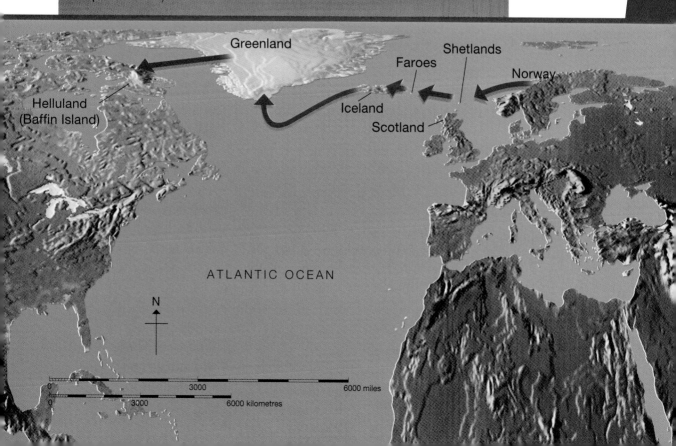

First people of St Lucia

St Lucia is an island country in the West Indies, to the east of the Caribbean Sea. Today about 150,000 people live on the island. The first people to live there were the Arawaks and Caribs, who sailed from South America. During the 17th century, the French and the British arrived from Europe. Control of the island changed hands many times between these two countries until 1814, when it was taken over officially by Britain.

Freed slaves

By 1814 the Europeans had brought many Africans to work as slaves on sugar **plantations**. Britain banned slavery throughout its empire 20 years later. Over the next 4 years 13,000 black slaves were freed. They were joined over the next 50 years by people from Portugal and India, who came to work on the island's plantations. The British gradually allowed the islanders more control and St Lucia became an independent country in 1979. Today's islanders are mostly **descendants** of African slaves.

● *St Lucia is a mountainous island with tropical vegetation. This map shows how its land is used. The capital, Castries, is the largest town and the main port.*

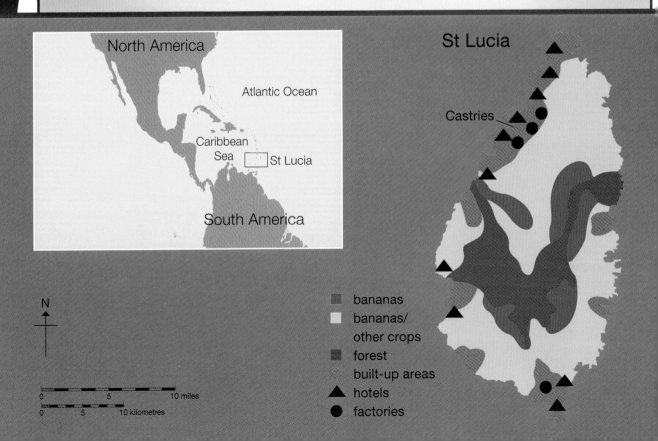

North America

Atlantic Ocean

Caribbean Sea — St Lucia

South America

St Lucia

Castries

N

	bananas
	bananas/ other crops
	forest
	built-up areas
▲	hotels
●	factories

0   5   10 miles
0   5   10 kilometres

Farming and industry

St Lucia's towns lie near the coast, but more than half the islanders live in rural areas. The island's main **resource** is bananas, which make up almost half of the country's **exports**. Many of the bananas are grown on small family farms, but there are also large plantations. The islanders also export coconuts and cocoa. There are a few factories producing clothes, electrical products, paper and textiles.

● *Holiday resorts on beautiful beaches like this one on the coast of St Lucia attract tourists from all over the world.*

Tourism

Like many Caribbean islands, St Lucia has a warm climate and lots of sandy beaches. This makes it attractive to holidaymakers from the USA, UK and many other parts of the world. It is a very hilly island, with **rainforest** and tropical **vegetation**. The **tourist** industry is growing fast, and this helps provide jobs for local people, especially in hotels and restaurants. Increased tourism also brings problems for many of the islanders. As more and more land is bought up by private companies, there is less room for small farms and local communities.

The importance of islands

Islands often have great importance for the countries to which they belong. They provide harbours, which help with trade, and offer coastlines and beaches that are attractive to **tourists**. Some islands have more **territory** than their **mainland** countries. The world's biggest island – Greenland – is a **self-governing province** of Denmark, which lies over 3000 kilometres (1864 miles) away in Europe. Greenland is about 50 times bigger than its mainland country but has a much smaller population.

Hong Kong

Hong Kong Island lies across Victoria Harbour from Kowloon peninsula, on the south-east coast of China. The island became British in 1842, after a war between China and the UK. It was agreed that the island would remain British for 99 years after 1898. Because of its natural harbour and its location in south-east Asia, Hong Kong grew as an important business and banking centre, as well as a busy port. In 1997 Hong Kong became part of China again.

⬤ *Hong Kong (which means 'fragrant harbour') is the name of an island and its main city as well as a region of China. The region is made up of more than 200 islands altogether.*

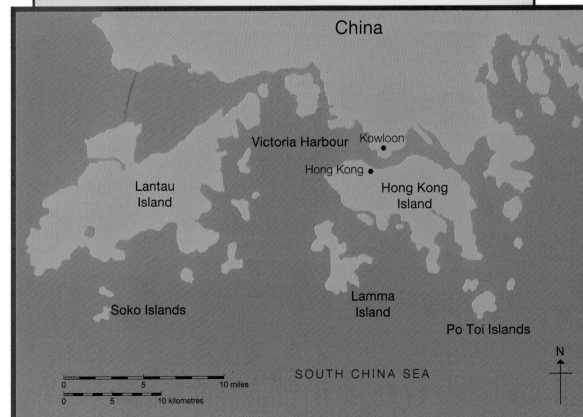

China

Victoria Harbour   Kowloon

Hong Kong

Lantau
Island

Hong Kong
Island

Soko Islands

Lamma
Island

Po Toi Islands

N

SOUTH CHINA SEA

0   5   10 miles
0   5   10 kilometres

New York

The largest city in the USA, New York, is built on islands. It is made up of five sections, called boroughs. Only one of these, the Bronx, is on the mainland. The boroughs of Brooklyn and Queens are part of Long Island. Staten Island and Manhattan are separate islands. There are many smaller islands around the city, including Liberty Island in New York Bay. This is where the Statue of Liberty stands, welcoming visitors to the city. Manhattan is New York's business and entertainment centre. Bridges and tunnels connect the island to the rest of the city.

● *Santa Cruz de Tenerife, in the Canary Islands, is an important deep-water port. The islands were named after the Latin for 'dog' (canis), because ancient sailors saw fierce dogs there. Canary birds are named after the islands.*

The Canary Islands

The 13 islands that make up the Canary Islands form 2 of the 50 provinces of Spain – but their nearest mainland is the African country of Morocco. These islands were important because they provided harbours when sailing south. Christopher Columbus stopped there to take on water and supplies before sailing across the Atlantic to the New World in 1492. The Islands' use has changed over the years. The Canaries have pleasant weather all year round, and today they are very popular with holidaymakers.

Changing islands

Islands change over long periods of time. **Continental** islands move further away from the **mainland**. Oceanic islands get lower when their **volcanoes** stop **erupting**. We never notice these tiny changes – but what if sea levels were to rise around the world? That would flood many coasts and have a terrible effect on low-lying islands.

Many scientists believe that ice at the North and South Poles is melting, because the worldwide temperature is rising. This is partly due to the **greenhouse effect**, which is caused by heat-trapping gases. These allow heat through to Earth from the Sun, but trap it there, like the glass of a greenhouse. The gases we release by burning **fuel**, such as the petrol we use to power our cars, add to this effect. It may cause a general increase in temperature, called **global warming**, which would create many problems – especially for islands.

● *This is one of the many islands that make up the Maldives. It has a small harbour for fishing boats and visiting yachts. The whole country has a population of about 285,000.*

The Maldives

The country of the Maldives is made up of about 1200 small **coral** islands that form a long **archipelago**. This chain of islands in the Indian Ocean is more than 750 kilometres (466 miles) long. Most of the islands are very low-lying, rising less than 2 metres above sea level. The highest point is just 24 metres high, on Wilingili Island. Coconut palms and fruit trees grow on the islands, many of which have become popular with **tourists**. Most of the islanders' income comes from fishing and tourism. Any rise in sea level would drown some of the islands, and a big rise would cover all of them.

The Queen of the Adriatic

Venice, in northern Italy, is an island city, built on about 120 **islets**. These are linked by more than 400 bridges. The city has canals instead of roads, so people travel by boat instead of by car. Its famous traditional boats are called gondolas (right). Founded in the 7th century, Venice became an important trading city, which some people called the 'Queen of the Adriatic'. It has many beautiful buildings and is a wonderful place for tourists to visit. The city is often flooded, however, and the floods have weakened the foundations of many buildings. This problem would be made worse by any rise in sea level.

Tourism

Islands are very popular with **tourists**. This is mainly because many islands have lots of beaches. When tourists first come to an island, the locals generally welcome them. Tourism brings jobs and income to the islanders – but when there are too many tourists, problems can arise. Small, quiet fishing villages turn into noisy places full of hotels and restaurants. Before too long, the original reasons people wanted to visit the islands are gone, and many locals move away.

Mediterranean islands

Many islands in the Mediterranean Sea have become very popular with tourists. These include the Spanish Balearic Islands. The Balearics are made up of four main islands, three of which have international airports. The largest island, Mallorca, is a popular destination for holidays. It has a few very busy beach resorts, but much of the island remains undeveloped. Next comes Menorca. Then Ibiza, which is well known as a fun place for young people to visit. The smallest island is Formentera, which is largely undeveloped.

● *In ancient times, the Balearic Islands were occupied by the Vandals, Phoenicians, Romans and many others. The islands became a Spanish province (part of Spain) in 1833. The capital of the province is Palma de Mallorca.*

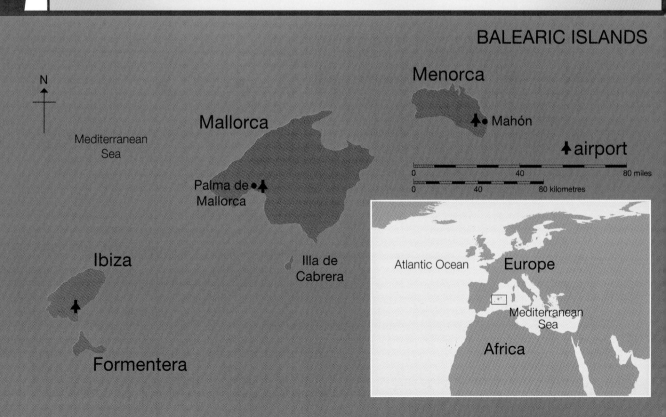

BALEARIC ISLANDS

Menorca

Mallorca

N

Mediterranean Sea

Mahón

airport

Palma de Mallorca

0 40 80 miles
0 40 80 kilometres

Ibiza

Illa de Cabrera

Atlantic Ocean   Europe

Mediterranean Sea

Africa

Formentera

Menorca

Over the past 20 years, the number of visitors to Menorca has grown dramatically. The island has a lot to offer: a large, deep-water harbour and an interesting town at Mahon, the capital. It also has wonderful swimming and boating off beaches all around the island. The Menorcans realized that their island would be ruined if it lost its charm. So, they stopped building high-rise hotels and said that all new villas must be built in a traditional style.

Biosphere Reserve

In 1993 UNESCO (the United Nations Educational, Scientific and Cultural Organization) declared Menorca a Biosphere Reserve. This means it is a protected island. The Menorcans have worked hard to look after the environmental features of their island. A nature reserve was created at the eastern end of the island.

● *These newly built villas surround a cove on the island of Menorca.*

Looking to the future

As we have seen, the world's islands change naturally over long periods of time. In recent years, many islands have also changed dramatically because of the way in which people have treated them. Some of these changes have been good, but others have spoiled islands' beauty. Today, nature **reserves** and national parks are often created to try and prevent unnecessary human development. The hope is that this will allow our islands to go on changing naturally in the future.

Island facts and figures

The world's biggest islands

island	country	area in sq km	area in sq miles
Greenland	Denmark	2,175,600	840,030
New Guinea	Indonesia, Papua New Guinea	821,030	317,001
Borneo	Indonesia, Malaysia, Brunei	744,366	287,401
Madagascar	Madagascar	587,041	226,657
Baffin Island	Canada	476,068	183,810
Sumatra	Indonesia	473,607	182,860
Honshu	Japan	230,448	88,976
Great Britain	United Kingdom	218,041	84,186
Ellesmere Island	Canada	212,688	82,119
Victoria Island	Canada	212,198	81,930

Islands with the greatest population

island	country	population in millions
Java	Indonesia	120
Honshu	Japan	101
Great Britain	UK	56
Sumatra	Indonesia	42
Luzon	Philippines	35
Taiwan	Taiwan	22
Sri Lanka	Sri Lanka	19
Hispaniola	Haiti, Dominican Republic	16
Mindanao	Philippines	15
Madagascar	Madagascar	14

The world's highest islands

island	country	highest point	height in m
New Guinea	Indonesia	Puncak Jaya	5030
Hawaii	USA	Mauna Kea	4205
Borneo	Malaysia	Kinabalu	4101
Taiwan	Taiwan	Yu Shan	3997
Sumatra	Indonesia	Kerinci	3806
Ross	Antarctica	Erebus	3794
Honshu	Japan	Fuji	3776
South Island	New Zealand	Cook (Aorangi)	3754
Lombok	Indonesia	Rinjahi	3726
Tenerife	Spain	Pico de Teide	3718

Biggest islands in each continent

island	continent	country	area in sq km	area in sq miles
Greenland	N. America	Denmark	2,175,600	840,003
Borneo	Asia	Indonesia, Malaysia, Brunei	744,366	287,401
Madagascar	Africa	Madagascar	578,041	223,182
Great Britain	Europe	UK	218,041	84,186
South Island	Oceania	New Zealand	150,737	58,199
Tierra del Fuego	S. America	Chile, Argentina	47,992	18,529
Alexander	Antarctica	–	43,200	16,679

The world's most densely populated islands

island	city/country	area in sq km	pop in mills*	pop per sq km	pop per sq mile
Manhattan	New York, USA	57	1.529	26,825	69,500
Salsette	Bombay, India	637	13.0	20,408	53,061
Hong Kong	China	75	1.33	17,733	47,500
Singapore	Singapore	572	3.476	6,077	15,800
Montreal	Montreal, Canada	471	1.8	3,822	9,945
Long Island	USA	4,406	7.4	1,679	4,350
Dakhin	Bangladesh	1,590	1.7	1,069	2,773
Java	Indonesia	132,186	120.0	908	2,351
Okinawa	Japan	1,434	1.2	907	2,170
Honshu	Japan	230,448	101.0	438	1,135

*Only including islands with a population of more than a million.

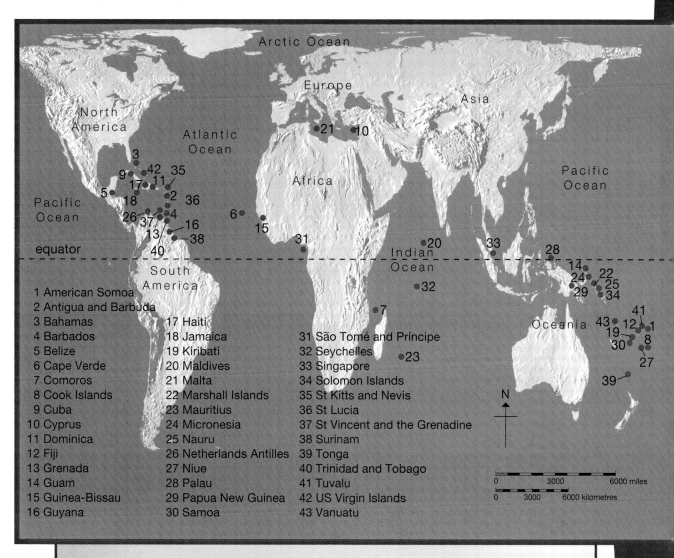

1 American Somoa
2 Antigua and Barbuda
3 Bahamas
4 Barbados
5 Belize
6 Cape Verde
7 Comoros
8 Cook Islands
9 Cuba
10 Cyprus
11 Dominica
12 Fiji
13 Grenada
14 Guam
15 Guinea-Bissau
16 Guyana
17 Haiti
18 Jamaica
19 Kiribati
20 Maldives
21 Malta
22 Marshall Islands
23 Mauritius
24 Micronesia
25 Nauru
26 Netherlands Antilles
27 Niue
28 Palau
29 Papua New Guinea
30 Samoa
31 São Tomé and Príncipe
32 Seychelles
33 Singapore
34 Solomon Islands
35 St Kitts and Nevis
36 St Lucia
37 St Vincent and the Grenadine
38 Surinam
39 Tonga
40 Trinidad and Tobago
41 Tuvalu
42 US Virgin Islands
43 Vanuatu

● *This map shows the location of members of the Small Island Developing States network (SIDSnet). This group of independent islands and low-lying coastal countries share concerns about their environment and try to work together within the United Nations.*

Glossary

anemones sea animal with stinging tentacles

archipelago group of islands

atoll ring-shaped group of coral islands or coral reef

barrier reef

civilization advanced culture, or a people with an advanced culture

colony large group of animals, plants or people that live close together

continent one of the world's seven huge land masses

continental relating to a large land mass (or continent)

coral tiny sea animal that forms colonies (called coral reefs) in warm, shallow waters

crust hard outer layer of Earth

descendant person who is born into a certain group of people

erupt (of a volcano) to throw out molten rock, ash and steam with an explosion

exports goods that are sent for sale to another country

extinct dead; an extinct volcano no longer erupts

fuel material that can be used as a source of energy, such as oil

glacier slowly moving mass of ice

global warming rise in temperature all over the world

greenhouse effect trapping of the Sun's warmth near Earth by gases in the air

hot spot very hot part of the layer of molten rock beneath Earth's crust; hot spots form volcanic islands

ice age period in the past when Earth was colder and more covered in ice

ice sheet thick layer of ice covering land

inhabit live on, or in

inlet small arm of the sea

islet small island

lagoon stretch of water surrounded and enclosed by coral islands or a reef

lava hot, molten rock that pours out of a volcano on to Earth's surface

mainland large piece of land, near which there may be smaller islands

molten melted (turned into liquid by heat)

nuclear explosion powerful, destructive release of energy caused by an atomic bomb

plantation farming estate where plants such as sugar cane or cotton are grown

plate huge piece of Earth's crust.

predator animal that hunts and kills other animals for food

prey animal that is hunted and eaten by another

rainforest thick forest found in warm, tropical areas with heavy rainfall

reef ridge of coral near the surface of the sea

reserve area set aside to protect the animals and plants that live there

resource supply of something that can be used

ridge raised strip of land, sand or other material

sandbank deposit of sand in shallow water

sand dune mound or hill of sand formed by the wind

seabed ground under the sea; the ocean floor

self-governing province region of a country that controls itself

species kind or type of animal or plant

territory land controlled by a country

tide rise and fall of sea level, which usually happens twice a day

tourist person visiting a place on holiday

unique one of a kind – only one

vegetation plants; plant life

volcano opening where molten rock and gas come from deep inside Earth, often forming a mountain

Further reading

Coasts and Islands, Terry Jennings (Belitha Press, 1999)
Earth Files: Islands, Chris Oxlade (Heinemann Library, 2003)
Mapping Earthforms: Islands, Catherine Chambers (Heinemann Library, 2000)

Index

Titles in the Landscapes and People series include:

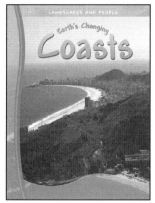

Hardback   1 844 21392 7

Hardback   1 844 21393 5

Hardback   1 844 21394 3

Hardback   1 844 21395 1

Hardback   1 844 21396 X

Hardback   1 844 21397 8

Find out about the other titles in this series on our website www.raintreepublishers.co.uk